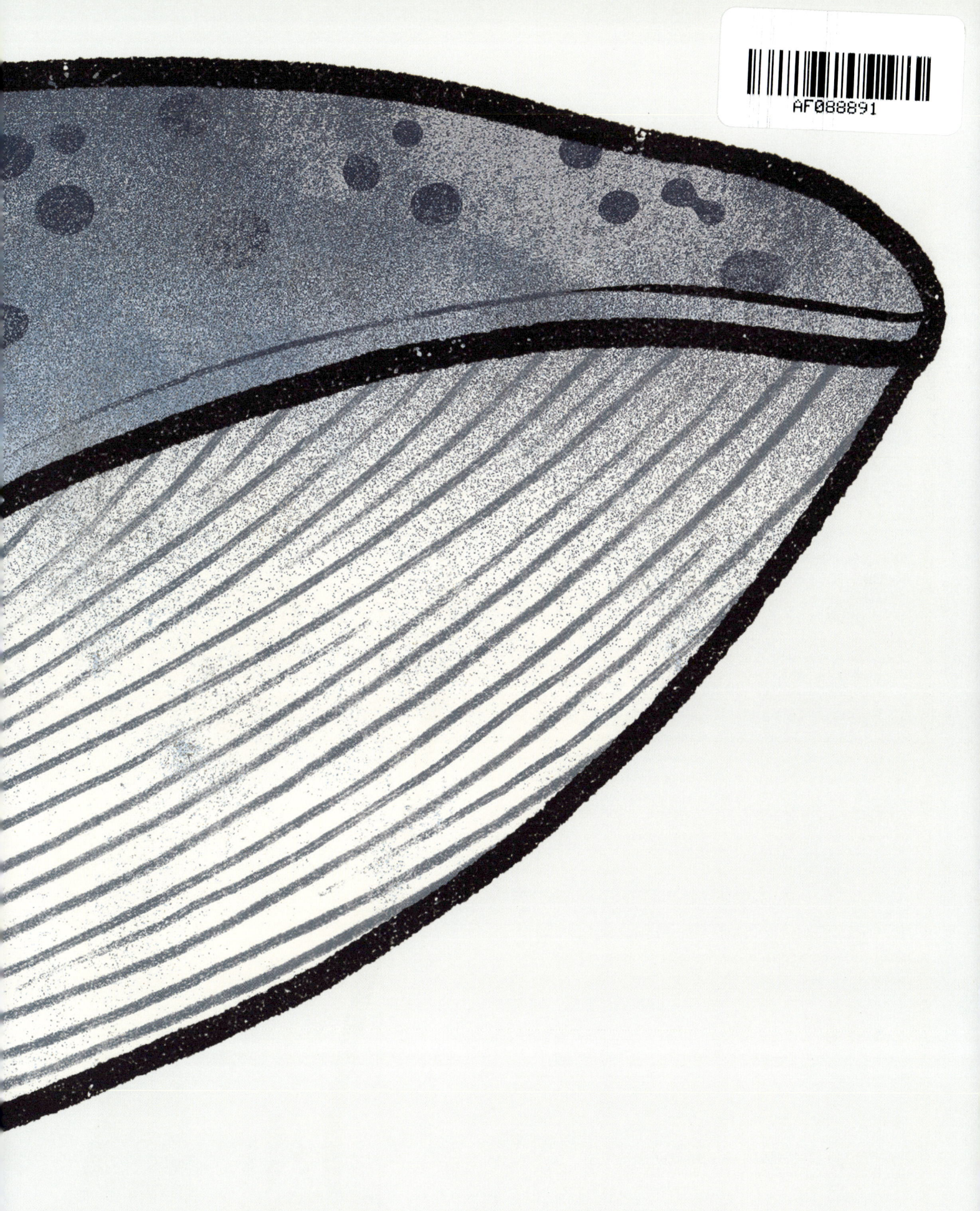

BOXER BOOKS Ltd. and the distinctive Boxer Books logo
are trademarks of Union Square & Co., LLC.
Union Square & Co., LLC, is a subsidiary of
Sterling Publishing Co., Inc.

Text © 2024 Boxer Books
Illustrations © 2024 Lo Cole

All rights reserved. No part of this publication may be reproduced,
stored in a retrieval system, or transmitted in any form or by any
means (including electronic, mechanical, photocopying, recording,
or otherwise) without prior written permission from the publisher.

First published in Great Britain in 2024.

ISBN 978-1-4547-1170-4

A catalogue record for this book is available from the British Library.

For information about custom editions, special sales,
and premium purchases, please contact
specialsales@unionsquareandco.com.

Printed in China
10 9 8 7 6 5 4 3 2 1

03/24

unionsquareandco.com

Spring Street™ Series created by David Bennett
Written by Sasha Morton
Illustrated by Lo Cole
Series editors: Sasha Morton and Leilani Sparrow
Series consultant: Mary Anne Wolpert, Cambridge University

ANIMALS
Contents

- **Popular pets** 6
- **On the farm** 8
- **African safari** 10
- **Big cats** 12
- **Amazing mammals** 14
- **Primates** 16
- **Marsupials** 18
- **Rainforest animals** 20
- **Rodents** 22
- **Helpful insects** 24
- **Birds, big and small** 26
- **More things with wings** 28
- **Reptiles on land** 30
- **Reptiles in the water** 32
- **Amphibians** 34
- **Marine mammals** 36
- **Under the sea** 38
- **Animals of the Arctic** 40
- **Animals of Antarctica** 42
- **Endangered animals** 44

Spring Street

Popular pets

1. Dalmatians don't develop their black spots until they are a few weeks old.
2. Siamese cats come from Thailand and are very talkative.
3. American shorthair cats are great at catching mice.
4. Golden retrievers love swimming.
5. Dachshunds are often called sausage dogs. Can you tell why?
6. There are over 70 species of betta fish.
7. Angelfish can live for up to 10 years in an aquarium.

Dogs are the most popular pets we own, mostly because they become such good friends with their owners. Cats and fish are the next most popular, and there are lots of different types of each to choose from. **Which pet would you pick?**

Guinea pigs sleep between 4 and 6 hours a day.

8

Netherland Dwarf rabbits are one of the smallest breeds of rabbit.

9

These little but lovable pets are popular, too. They are easy to pick up and hold, but they still need lots of attention and exercise. Which small pet is your favourite?

10

Gerbils wash themselves with sand instead of water.

11 Pet mice can jump up to 12 inches (30cm) in the air.

Syrian hamsters are happy to live by themselves.

12

1	Dalmatian	4	Golden retriever	7	angelfish	10 gerbil
2	Siamese cat	5	Dachshund	8	guinea pig	11 mouse
3	American shorthair cat	6	betta fish	9	Netherland Dwarf rabbit	12 Syrian hamster

7

On the farm

1 There are more than 1 billion sheep on our planet.

2 Goats are scared of water.

3 Turkeys have 5,000 to 6,000 feathers.

4 Alpacas and llamas are both part of the camel family.

5

Did you know that groups of animals have different names? A group of horses, cows, sheep, goats, alpacas and llamas is a herd. Chickens and wild turkeys live in a flock, and a group of turkeys living on a farm is called a gaggle.

6 There are more than 22 billion chickens on our planet.

7 Cows have an amazing sense of smell.

8 Horses can sleep standing up.

- 1 sheep
- 2 goat
- 3 turkey
- 4 alpaca
- 5 llama
- 6 chickens
- 7 cow
- 8 horse

African safari

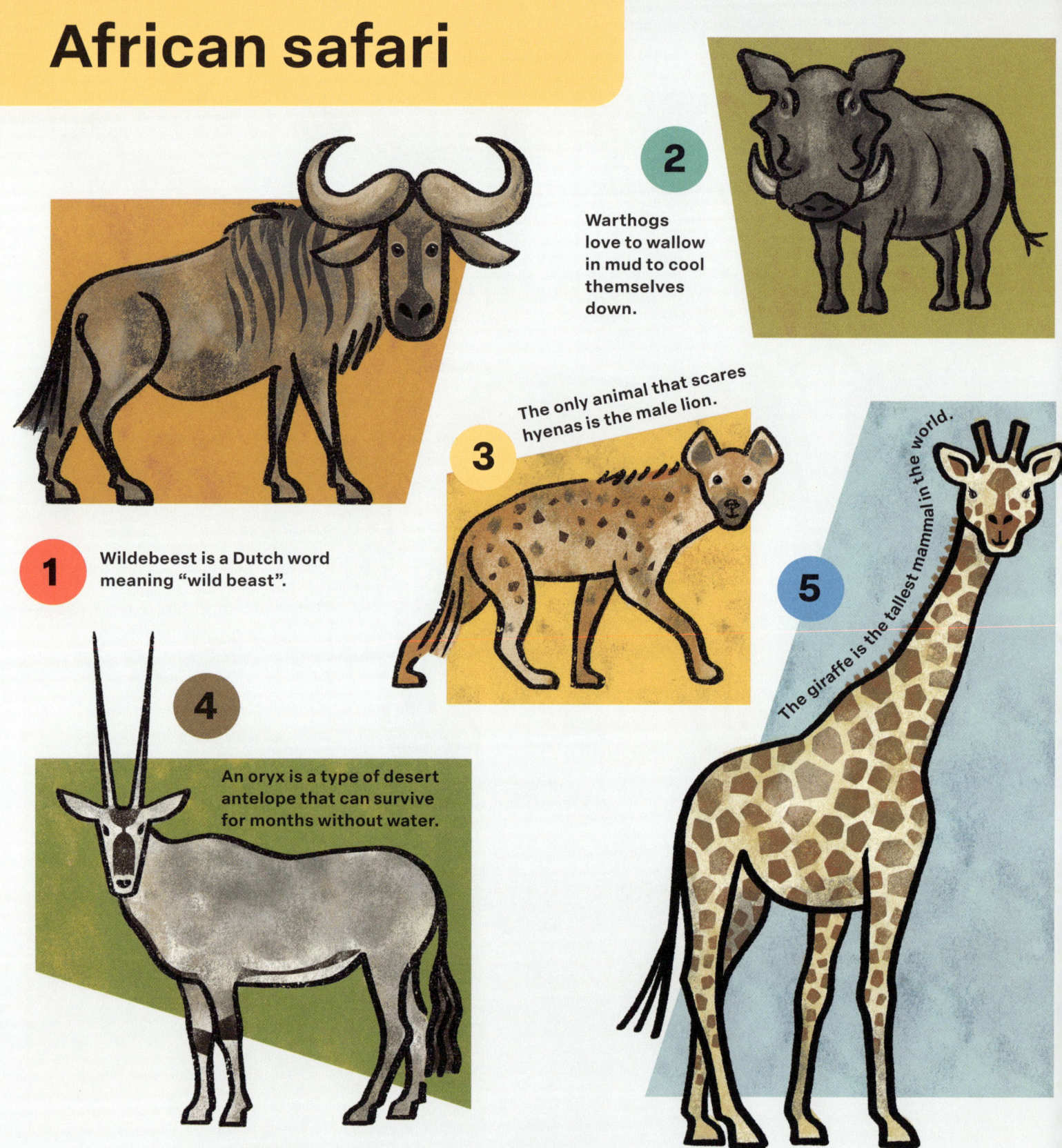

1 Wildebeest is a Dutch word meaning "wild beast".

2 Warthogs love to wallow in mud to cool themselves down.

3 The only animal that scares hyenas is the male lion.

4 An oryx is a type of desert antelope that can survive for months without water.

5 The giraffe is the tallest mammal in the world.

On an African safari, a guide drives people through protected grasslands, plains or deserts to look at Africa's wildlife in its natural habitat. **Which of these animals would you most want to see on safari?**

6

African bush elephants are the largest and heaviest land animals on Earth.

7

In Africa, caracals are known as "little lions".

A zebra's stripes help to camouflage it in tall grasses.

8

9

African or Cape buffalos can live in large herds with up to 2,000 members.

1	wildebeest	4	oryx	7	caracal
2	warthog	5	giraffe	8	zebra
3	hyena	6	elephant	9	African buffalo

Big cats

1 Lions hunt mostly at night.

2 Jaguars can bite through turtle shells and crocodile scales.

3 India has around 3,000 of the world's remaining wild tigers.

Lions, leopards, tigers, jaguars and snow leopards are part of a special group known as *Panthera*. These five big cats are found in various habitats around the world, from rainforests and deserts to mountains. The jaguar is the only type of big cat that lives in the Americas. Cheetahs and cougars are similar to the five big cats.

Amazing mammals

1 Water deer from China and Korea have long tusks instead of antlers.

2 A group of meerkats is called a mob.

3

Raccoons are nearly as intelligent as monkeys.

The suni is a very small antelope.

4

Any living creature born with a backbone and hair, and that drinks its mother's milk, is called a mammal. In fact, humans are mammals! Here are some of the world's small and interesting mammals.

5 Hedgehogs can climb, run and swim.

6 Badgers have strong front legs and long claws that make them very powerful diggers.

7 This is an Egyptian mongoose. The plural of "mongoose" is mongooses, not mongeese.

8 Fennec foxes use their huge ears to help them hunt at night.

9 The elephant shrew is one of the fastest small mammals in the world.

10 The pink fairy armadillo is only found in Argentina.

1. water deer
2. meerkats
3. raccoon
4. suni
5. hedgehog
6. badger
7. Egyptian mongoose
8. fennec fox
9. elephant shrew
10. pink fairy armadillo

Primates

1. Orangutans are part of a group called the Great Apes.
2. Chimpanzees live in central and west Africa.
3. Every gorilla's "noseprint" is unique.
4. Bonobo groups are usually led by females.

Primates are a group of mammals who have hands, feet and eyes that look forward. They are also extremely good at living in trees, and many make their homes in high branches.

5 As well as sounding like a baby crying, a bush baby can also croak, cluck and whistle.

6 Every ring-tailed lemur has 13 white stripes on its tail, with a black stripe on either side of it.

7 The tiny marmoset can live for up to 17 years.

8 Unlike most primates, which eat a mixture of plants, meat and insects, tarsiers only eat meat.

9 There are more than 260 different species of monkey.

10 The loris has a special layer across its eyes that gives it excellent night vision.

1 orangutan
2 chimpanzee
3 gorilla
4 bonobo
5 bush baby
6 ring-tailed lemur
7 marmoset
8 tarsier
9 monkey
10 loris

Marsupials

1. Kangaroos can leap more than 29 feet (9m) in a single bound.

2. Sugar gliders use the stretchy skin between their wrists and ankles like wings to let them glide from tree to tree.

3. Quokkas are known for their happy faces.

4. Yellow-footed rock-wallabies can jump as far as 13 feet (4m).

Marsupials are animals that still have some growing to do after they're born, so they live in a pouch on the front of their mother's body. If they get scared once they are out in the big, wide world, they scamper back to the pouch they grew up in.

Most marsupials live in Australia – the opossum is North America's only marsupial.

Koalas mostly eat eucalyptus leaves, which are poisonous to almost all other animals.

Wombats have special fur that can glow in the dark.

When a Tasmanian devil is born, it is the size of a raisin.

1	kangaroo	**3**	quokka	**5**	opossum	**7**	wombat
2	sugar glider	**4**	yellow-footed rock-wallabies	**6**	koalas	**8**	Tasmanian devil

Rainforest animals

1 Orangutans are able to eat with their feet.

2 Toucans lay their eggs in holes in tree trunks.

3 A chameleon's colour changes are a reaction to temperature, light and its mood.

4 Emerald tree boas give birth to live snakes instead of laying eggs that then hatch.

In the rainforest canopy, you'll find amazing creatures that spend nearly all their time living far above the forest floor. Further down, the forest floor is filled with colourful creatures.

The world's oldest parrot is a macaw called Charlie, who is believed to be around 125 years old!

5 Sloths are slow-moving on land but move three times faster in water.

7 Spider monkeys have strong tails but no thumbs.

8 The blue morpho butterfly's wings are covered in tiny scales that reflect the light.

9 Red-eyed tree frogs use their red eyes to frighten off predators.

1	orangutan	**4**	emerald tree boa	**7**	spider monkey
2	toucan	**5**	sloth	**8**	blue morpho butterfly
3	chameleon	**6**	parrot	**9**	red-eyed tree frog

Rodents

1. In the wild, hundreds of chinchillas can live in one colony.

2. Naked mole rats don't drink water.

3. Rats "laugh" when they are tickled.

4. Capybaras are the world's largest rodent.

5. Chipmunks are a type of tiny squirrel.

Rodents are mammals that have a pair of sharp front teeth in their upper and lower jaws that never stop growing. This is why they gnaw on things. It stops their teeth from growing too long! Some rodents, such as chinchillas, have become popular pets.

Helpful insects

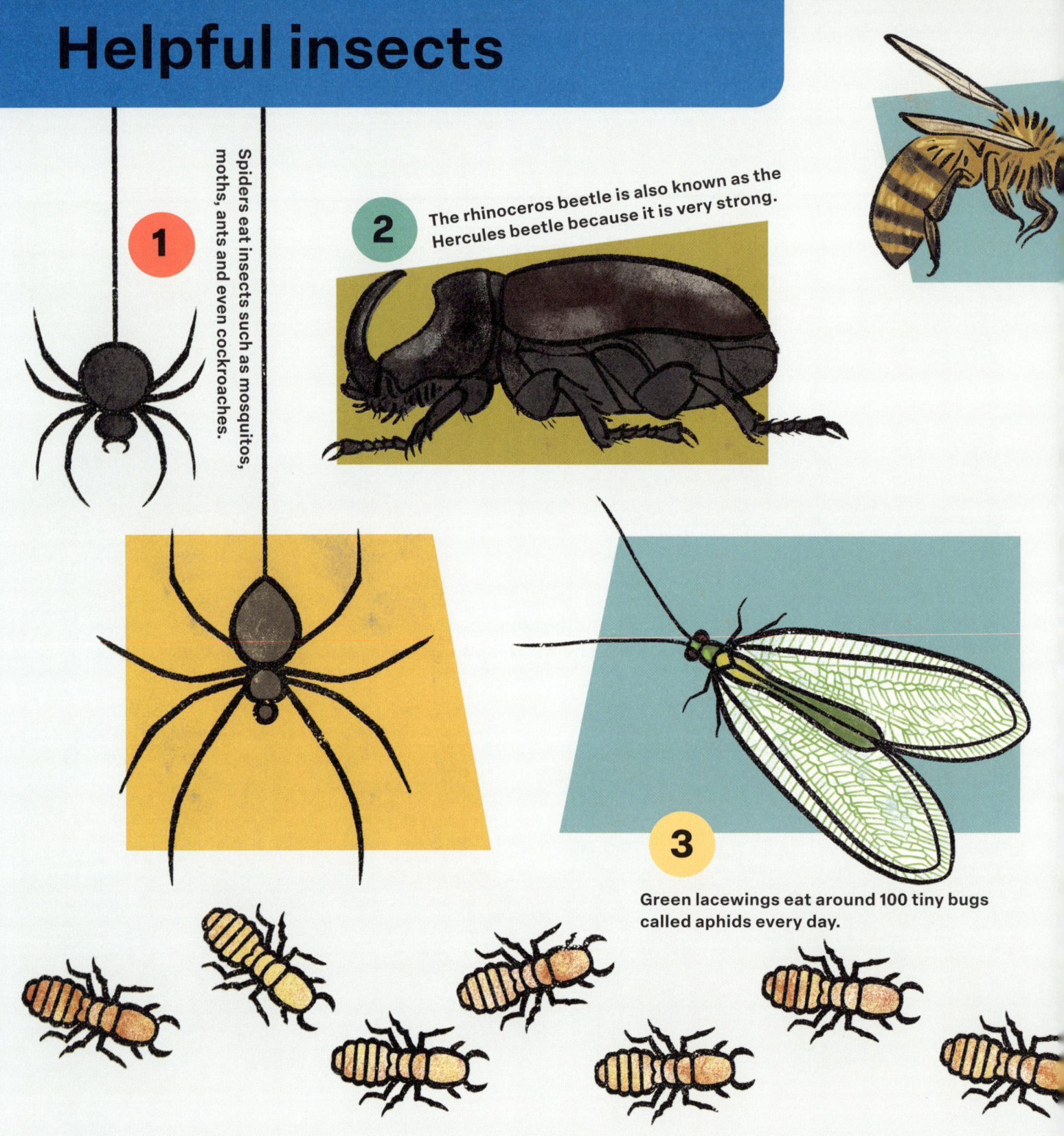

1 Spiders eat insects such as mosquitos, moths, ants and even cockroaches.

2 The rhinoceros beetle is also known as the Hercules beetle because it is very strong.

3 Green lacewings eat around 100 tiny bugs called aphids every day.

Bees are the perfect example of small creatures that are incredibly important for our planet. They transport pollen from plant to plant to help them grow, which includes crops that provide us with food. These plants help keep the air we breathe clean, too. In fact, all of these little creatures are helpful at removing things from their environment, which helps improve ours!

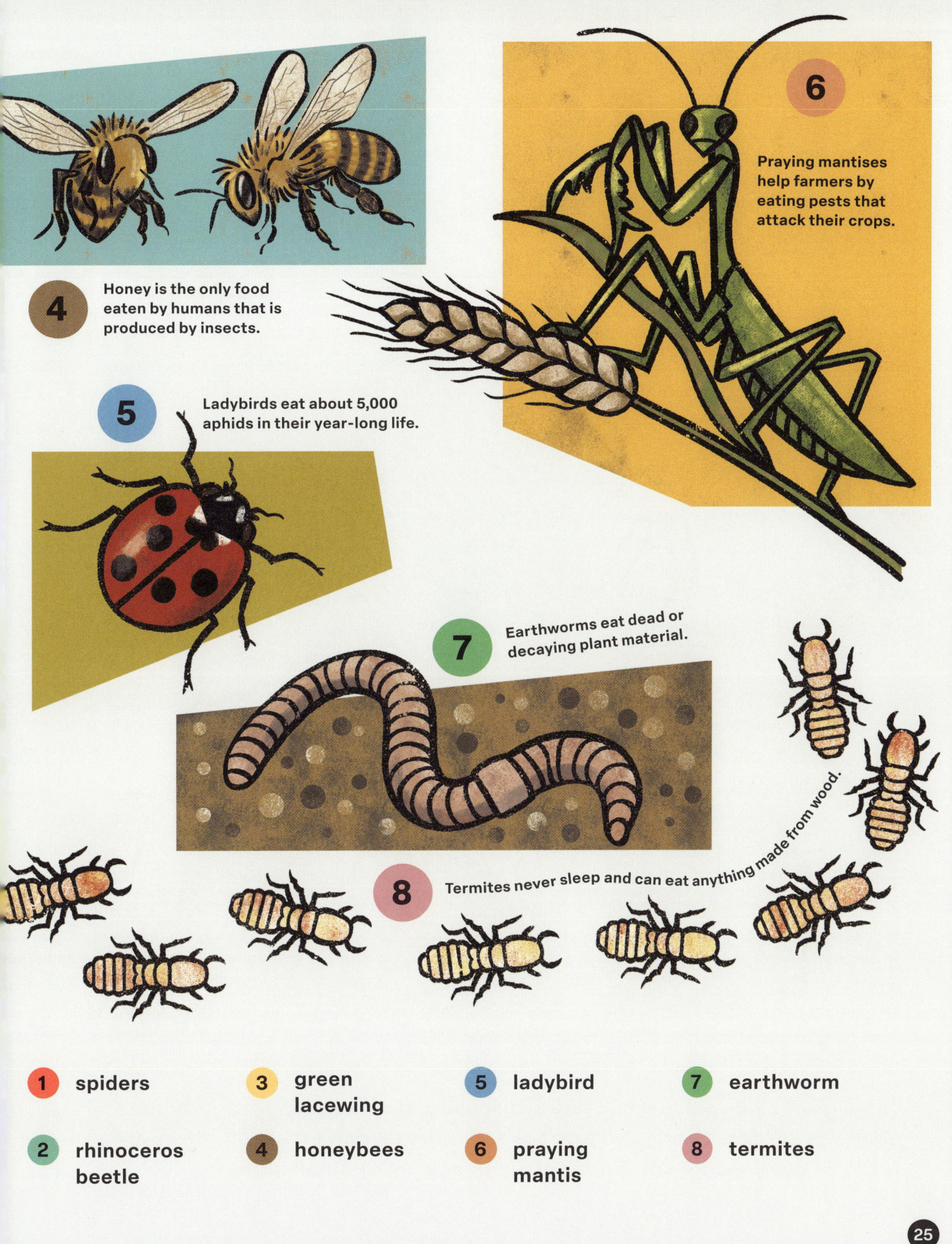

Birds, big and small

1 Woodpeckers drum on trees with their beaks as a way of communicating.

2 Because of their brightly coloured beaks, puffins are nicknamed "parrots of the sea".

3 Flamingos are believed to stand on one leg to keep warm.

4 The Philippine eagle has a wingspan of 6 feet (2m).

5 The pink robin lives in Australia and feeds on insects found on forest floors.

6 The bee hummingbird comes from Cuba and is the smallest bird in the world.

Birds can be big, small and every size in between. From birds of prey to seabirds, birds with huge wingspans to birds that can't fly at all, there are birds on every continent. **Have you seen any big or small birds today?**

When huge groups of starlings swoop and swirl through the sky, it's called a murmuration.

Gulls often steal food from other birds.

Mandarin ducks are very shy and like to nest in trees.

Kiwis, emus and ostriches are all flightless birds.

1 woodpecker	4 Philippine eagle	7 starlings	10 kiwi
2 puffin	5 pink robin	8 gull	11 emu
3 flamingo	6 bee hummingbird	9 mandarin duck	12 ostrich

More things with wings

2 Peppered moths are one of the world's most common moths.

3 The housefly uses its sense of smell to find food.

1 Flying squirrels use their fluffy tails to stabilise themselves in flight and to put on the brakes.

4 Dragonflies hunt other flying insects and can fly backwards.

5 Monarch butterflies can fly between 50 to 100 miles (80 to 160km) every day.

Did you know there are around 160,000 different kinds of moths, 17,000 different butterflies and 1,400 different bats flying across every continent except Antarctica? Here are just a few of them and some other things with wings!

6 Bats are actually mammals.

7 Wasps are attracted to bright colours.

8 Adult luna moths don't have mouths.

9 Western pygmy blue butterflies are only around ½ inch (13mm) across.

10 The orange oakleaf butterfly can disguise itself as a leaf when it closes its wings.

1 flying squirrel
2 peppered moth
3 housefly
4 dragonfly
5 monarch butterfly
6 bat
7 wasp
8 luna moth
9 Western pygmy blue butterfly
10 orange oakleaf butterfly

Reptiles on land

1 Komodo dragons are the largest species of monitor lizard in the world.

2 Other than snakes, monitor lizards are the only reptile that has a forked tongue.

Reptiles are among the longest-living things that have existed on our planet – and dinosaurs were reptiles, too! These creatures have dry scaly skin, and most of them lay eggs that their young hatch from.

3 Agama lizards enjoy basking in the sun.

4 Chameleons can change colour to blend into their surroundings.

5 Pythons swallow their prey whole.

6 Corn snakes are usually active at night.

7 Geckos can hang by their toe hairs and regrow their tails.

1 Komodo dragon
2 monitor lizard
3 agama lizard
4 chameleon
5 python
6 corn snake
7 gecko

31

Reptiles in the water

1 Saltwater crocodiles have their eyes, ears and nostrils on top of their heads.

2 Olive-headed sea snakes usually live in coral reefs.

3 Leatherback sea turtles are the largest type of sea turtle and have existed for over 100 million years.

4

There are four groups of marine reptiles that live in the salt water that makes up our planet's seas and oceans: sea turtles, sea snakes, marine iguanas and saltwater crocodiles.

Marine iguanas are only found in the Galápagos Islands.

Alligators can't survive in salt water. They hang out in freshwater rivers, marshes, swamps and lakes, with more than 1.5 million alligators living in the state of Florida.

A yellow-lipped sea krait is very venomous.

A female green sea turtle can lay more than 100 eggs.

1. saltwater crocodile
2. olive-headed sea snake
3. leatherback sea turtle
4. yellow-lipped sea krait
5. marine iguana
6. alligator
7. green sea turtle

Amphibians

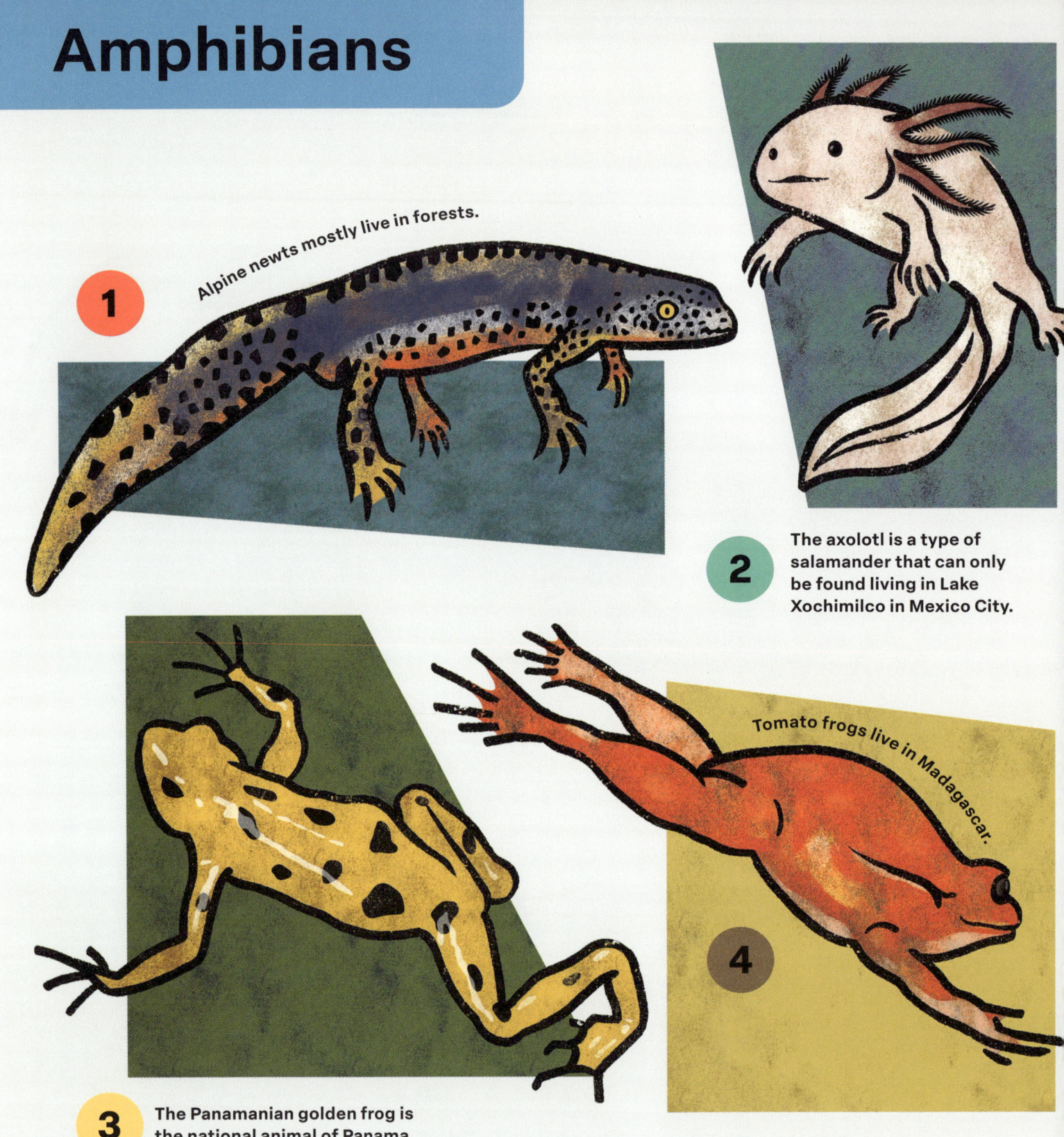

1 Alpine newts mostly live in forests.

2 The axolotl is a type of salamander that can only be found living in Lake Xochimilco in Mexico City.

3 The Panamanian golden frog is the national animal of Panama.

4 Tomato frogs live in Madagascar.

Amphibians, such as frogs and toads, are cold-blooded creatures with backbones, and they can live on both land and water. They must breed and develop into adults in water, and while they don't have scales, most of them can breathe through their skin or gills.

5 Caecilians are snakelike and hide in soil or streambeds.

6 Salamanders can grow a new limb if they lose one.

7 The natterjack toad is sometimes called the running toad because it crawls quickly instead of hopping.

1. alpine newt
2. axolotl
3. Panamanian golden frog
4. tomato frog
5. caecilian
6. salamander
7. natterjack toad

Marine mammals

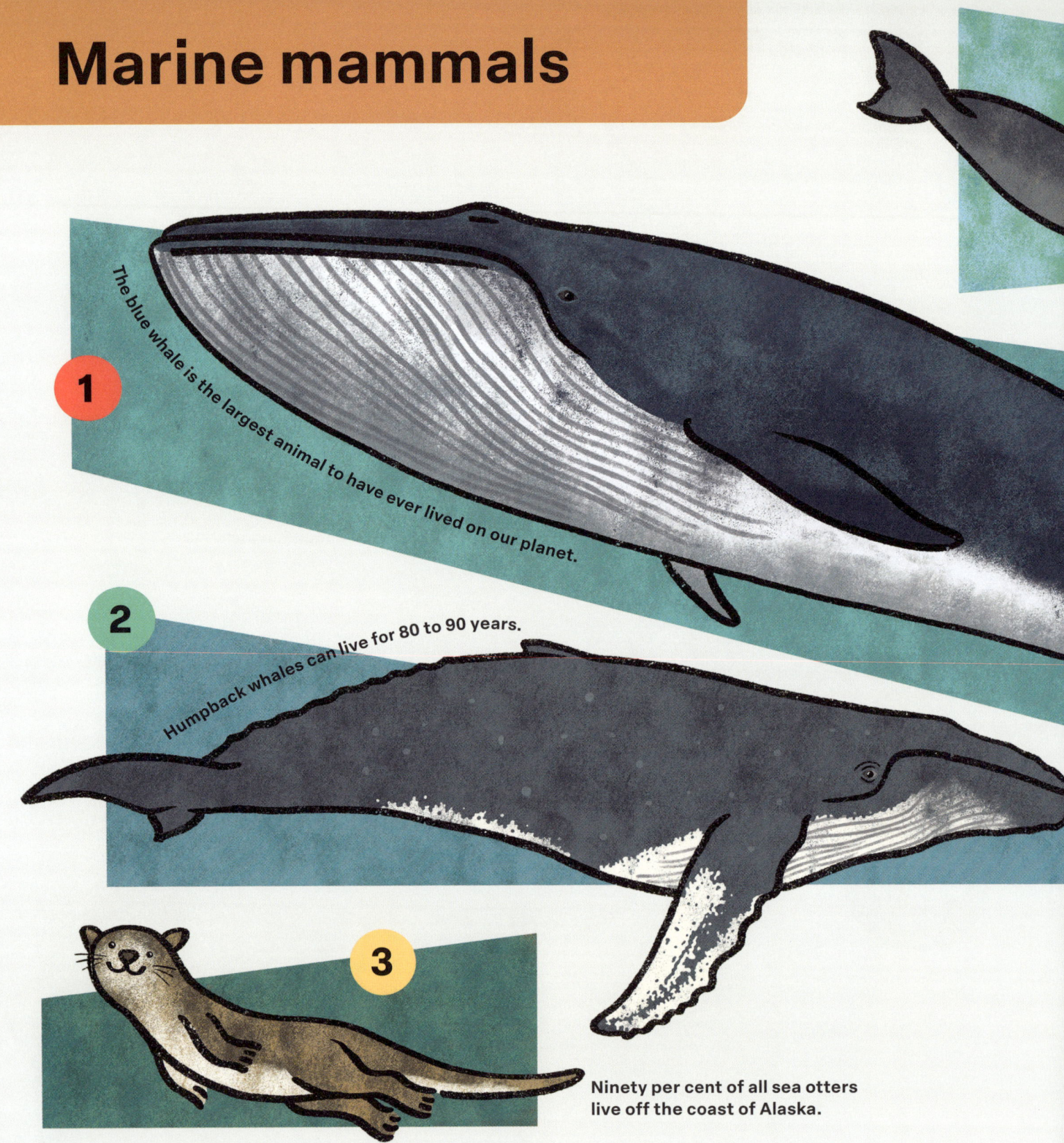

1. The blue whale is the largest animal to have ever lived on our planet.

2. Humpback whales can live for 80 to 90 years.

3. Ninety per cent of all sea otters live off the coast of Alaska.

The five oceans cover more than 70 per cent of the Earth's surface and are home to around 90 per cent of all living things on our planet. Some of the smallest creatures, such as zooplankton (which can only be seen under a microscope), live alongside the largest animal ever to live on Earth: the blue whale.

Under the sea

Great white sharks have around 300 very sharp teeth.

1

2 Japanese flying squid can actually fly.

3 Female anglerfish have a glowing light above their heads.

4 Before dinosaurs walked the earth, there were jellyfish on our planet.

Some of the world's most weird and wonderful living things live under the sea! Deep beneath the water's surface, you can find fantastic fins, tangly tentacles and creatures that glow in the dark.

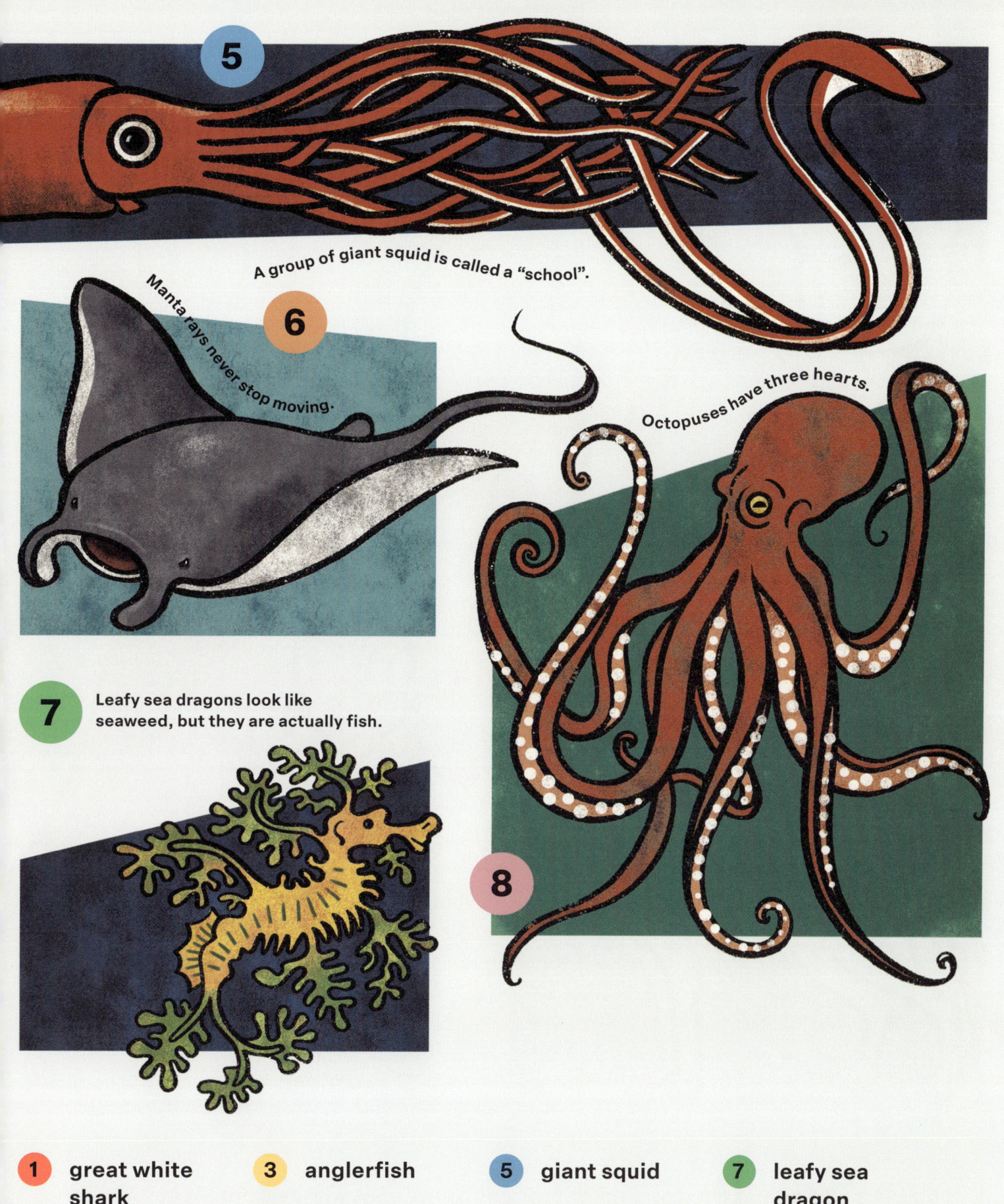

5 A group of giant squid is called a "school".

6 Manta rays never stop moving.

7 Leafy sea dragons look like seaweed, but they are actually fish.

8 Octopuses have three hearts.

1	great white shark	**3**	anglerfish	**5**	giant squid	**7**	leafy sea dragon
2	Japanese flying squid	**4**	jellyfish	**6**	manta ray	**8**	octopus

Animals of the Arctic

1 Polar bears are the largest type of bear in the world.

2 Arctic hares have short ears so there is less surface area to get cold.

3 The Arctic wolf has an outer layer of waterproof and snow-proof fur.

4 Arctic foxes have white fur in the winter that changes to brown-grey when spring comes.

The land and sea at the top and bottom of our planet are extremely cold. Check out these clever creatures that have adapted to live in freezing places.

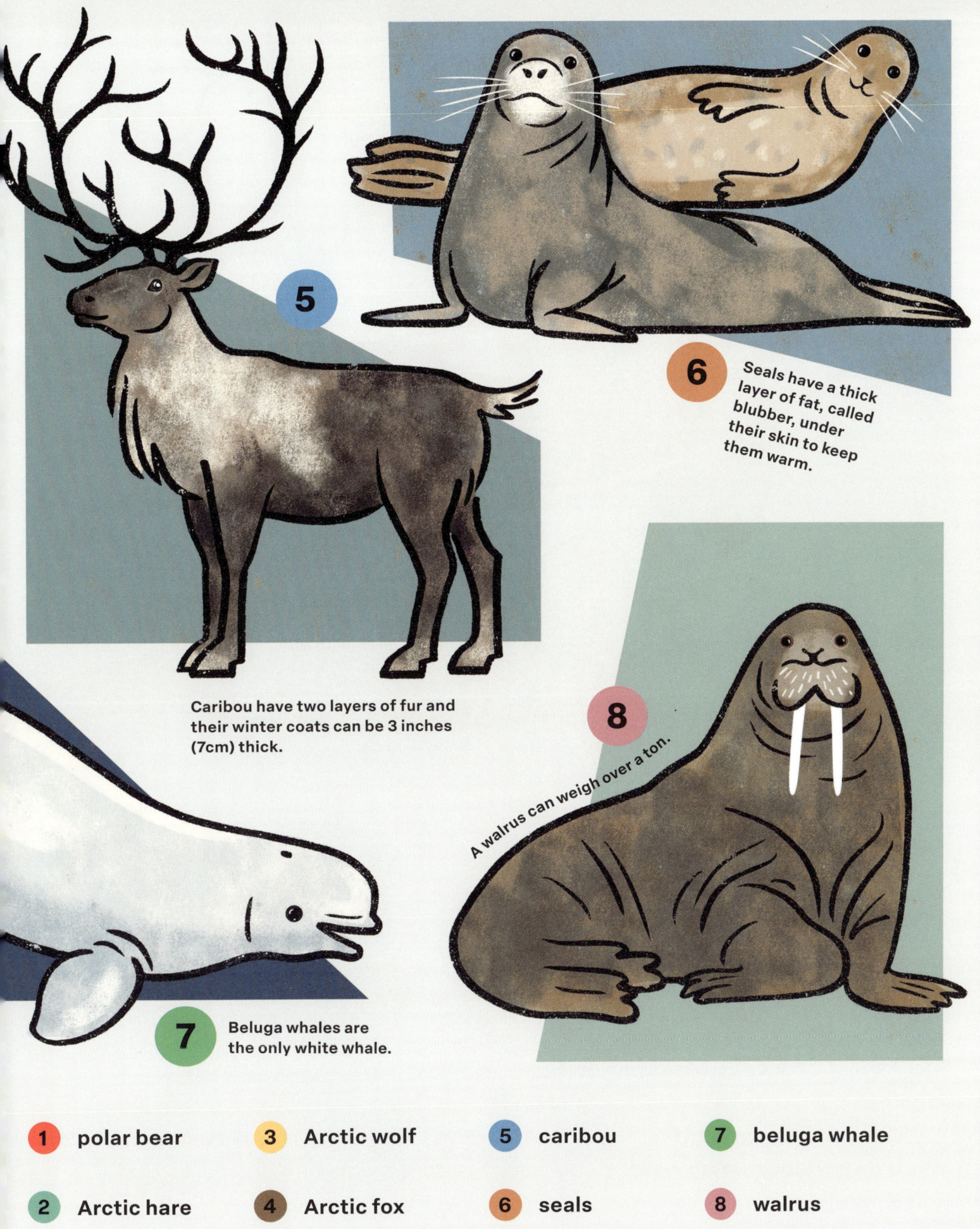

5 Caribou have two layers of fur and their winter coats can be 3 inches (7cm) thick.

6 Seals have a thick layer of fat, called blubber, under their skin to keep them warm.

7 Beluga whales are the only white whale.

8 A walrus can weigh over a ton.

1. polar bear
2. Arctic hare
3. Arctic wolf
4. Arctic fox
5. caribou
6. seals
7. beluga whale
8. walrus

Animals of Antarctica

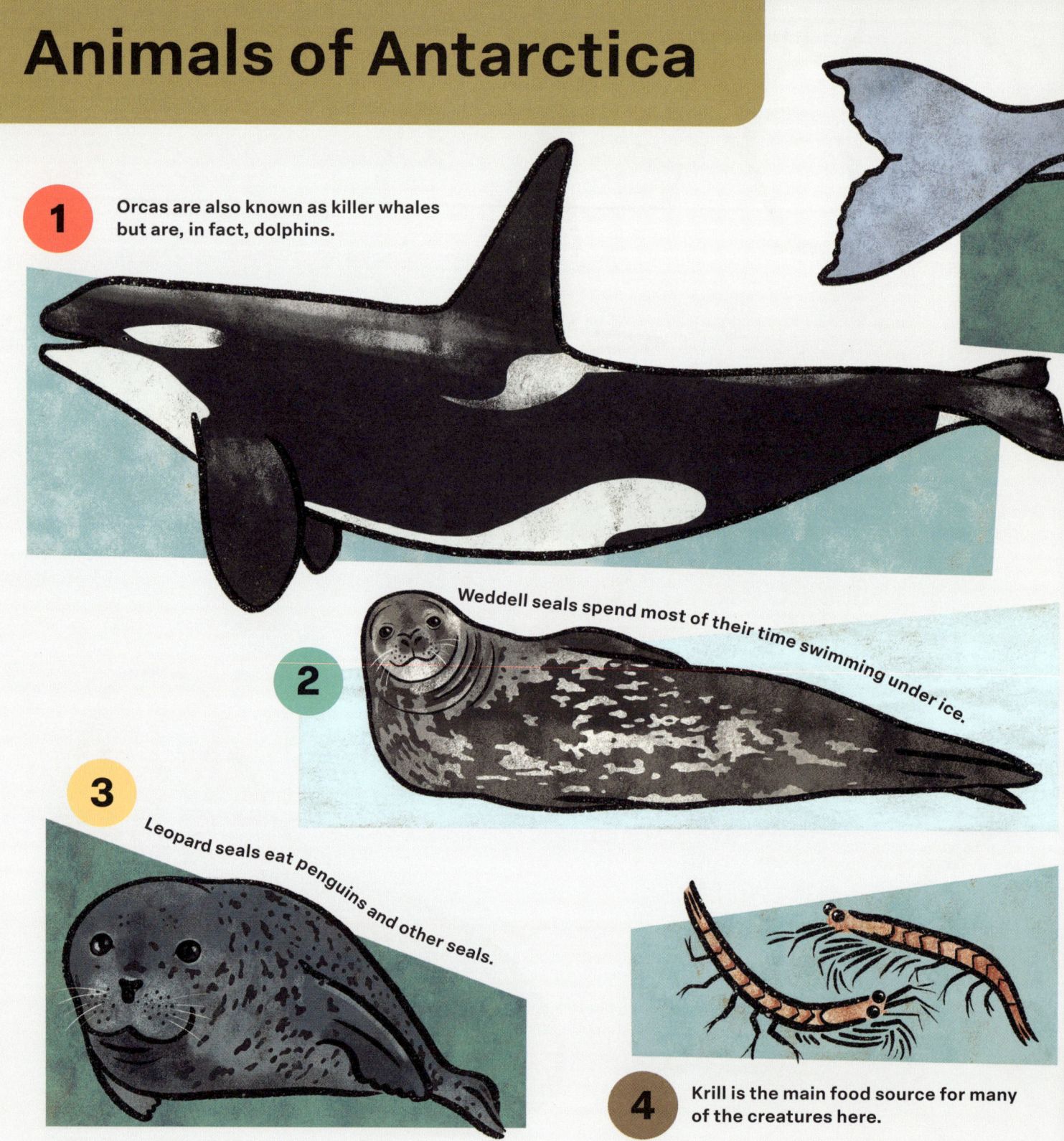

1 Orcas are also known as killer whales but are, in fact, dolphins.

2 Weddell seals spend most of their time swimming under ice.

3 Leopard seals eat penguins and other seals.

4 Krill is the main food source for many of the creatures here.

No humans can make their home in Antarctica all the time because it's too cold. You'll need to wrap up warm if you want to see the amazing creatures that live here!

5 Baleen whales, such as blue or humpback whales, use bristles instead of teeth to catch their food.

6 The albatross has the largest wingspan of any bird, up to 11 feet (3.3m).

7 Adélie penguins are the smallest penguins living in Antarctica.

8 Southern rockhopper penguins use their webbed feet to steer when swimming underwater.

9 Emperor penguins are the largest of all the penguins.

1. orca
2. Weddell seal
3. leopard seal
4. krill
5. baleen whale
6. albatross
7. Adélie penguins
8. southern rockhopper penguin
9. emperor penguin

Endangered animals

1 The African forest elephant is found in west and central Africa.

2 Hawksbill turtles are found in the waters of the Atlantic, Indian and Pacific oceans.

3 The Sunda Island tiger is found in Sumatra, Indonesia.

4 There are about 100 Amur leopards living in the wild. They live in forests in southeast Russia. Snow leopards are found in mountainous regions of central and south Asia.

As the amount of space for animals who share our planet gets smaller, their numbers get smaller, too. These are some of the creatures that may become extinct unless we take care of our planet and the animals living on it.

5 There are fewer than 100 Javan rhinoceroses left on the planet. Black rhinos of eastern and southern Africa are also very endangered.

6 Mountain gorillas are found in central and eastern Africa.

7 Yangtze finless porpoises are found in the Yangtze River in China.

8 Tapanuli and Sumatran orangutans are found in Sumatra, Indonesia.

1. African forest elephant
2. hawksbill turtle
3. Sunda Island tiger
4. Amur and snow leopards
5. Javan and black rhinos
6. mountain gorilla
7. Yangtze finless porpoise
8. Tapanuli and Sumatran orangutans